Don't Farm Naked

AND OTHER MUST HAVE TIPS TO CULTIVATE STUDENT SUCCESS AND CAREER HAPPINESS

DONNA FULLER

Acme and Roboto fonts used with permission from Google.

ISBN-13: 978-0-9984243-4-7

ISBN-13: 978-0-9984243-5-4

This book is dedicated to any student:

~You *might* be one of my people if... you're going to school to get somebody to stop nagging you.

~You *might* be one of my people if... you're picking a major because it requires the least amount of math, science, history, and government.

~You *might* be one of my people if...you're picking a school because it has the highest partying rate in the state.

~You *might* be one of my people if... you're going to school because you have absolutely nothing else better to do.

~You *might* be one of my people if... you're going to school because somebody said to.

~You *might* be one of my people if... you're picking a school because it's the one offering a major in beer brewing.

If you are one of my people,
this book is for you.

CONTENTS

Where do farmers send their kids to grow?

Kinder-garden

Introduction

For an industry that literally feeds the world, agricultural (ag) students and professionals sure are an overlooked and underserved population. I aim to change that because ag people are my people.

My farm story

My dad is a veteran, farmer, and was raised by cotton-pickin' farmers, so he has the work ethic of a workhorse. Growing up, I watched him work Monday through Friday for 40 hours a week at his "real job." Then, on Saturday and Sunday, the whole family got up early to work at the farm 'till sundown. I imagine my dad preferred tractor therapy over his "real job," but come Monday, he did it all over again.

All that is admirable, but there was no work-life balance of any kind. If it weren't for my mom, I'd have never seen the beach. To my dad, "fun" was a four-letter word.

Today, he is 85 years old. Yet, during our Texas summers, when we have blast furnace temperatures, he's haulin' hay! Thanks to my mom, at least he's in an air-conditioned tractor cab, but still...

If you're an ag person, you're probably thinking, "Um, yeah, what's your point?" This is because most of you were raised with a love for the land and a workhorse work ethic.

I love ag kiddos because they have thicker skin and work to earn their keep. They know to get what they want, they gotta give it some elbow grease. They were raised right.

Since being raised on a cattle farm, I now live in a barn-do-minium, on a goat farm, surrounded by all sorts of critters. However, a country bumpkin I am not. Okay, maybe a little bumpkin at times. After all, I'll still pee in the woods...

My college story

My parents had good intentions. They wanted me to have a great future with options and make money, so I wouldn't have to struggle through life. Therefore, I was expected to go to college. However, as high school graduation got closer, the struggle became real. Our differences intensified, and my desire to party magnified. Their good intentions became less about giving me a better life and more about getting me out of the house.

These feelings were mutual. And so, college seemed like a great escape. The problem was, nobody told me what to do before I got there. I didn't know my options. I didn't have any sort of real direction, and I couldn't see my future. Quickly, I became that student who felt lost, confused, and hopeless.

I was alone and lonely. From experience, I can tell you there is nothing more debilitating than having no direction or believing you're stuck doing something you don't want to do. As a consequence, to avoid feeling depressed, I made sure there was a lot more fun than studying going on. Fun, as in drinkin', eatin', and skippin' school. I drowned myself in free soft-serve ice cream in the college cafeteria to feel comforted, packing on the "freshman 20" or more. If we were talking dog years, that would be more like 120. And it felt like it.

My career story

Fast forward 100 years to my role as an academic advisor. Ag students and I connected quickly because of our similar upbringing. But after speaking with them, I realized they consistently had many of the same things in common:

- They were in school for the wrong reasons.
- They didn't know themselves nor their options.
- They sounded lost, confused, and hopeless.

Some students were more on track than others, but many were taking classes just to take classes, without understanding what education or career route was a good fit for them.

There were plenty of ag students who *claimed* they wanted to go to veterinary school. However, this is kind of hard when you've got a 310 on the math part of your SAT and are either unwilling to take or unable to pass an animal science class. Most students were clueless about the amount of time, money, and GPA required to get there. When I gave them the facts, they practically swallowed their gum.

Worst of all, they had no clear direction for their future. My heart broke when I spoke to them. Those feelings of loneliness, confusion, and despair came flooding back. I realized we had more in common than both being farming fans—I had once been one of them! No wonder we had such a strong connection!

My why

I aim to change things because:

- I have been in your boots, so I get you. I never want anyone to feel what I felt, so I empathize with you.
- I want to give you the support and attention you need so you have a clear

mind and feel secure coming out of the chute!

My good fit

I am a good fit for you because:

- I am completely invested in your future. However, I won't nag you into doing something you don't want to do.
- I will show you the ropes!
- I will help you discover your strengths, skills, and passions.
- I will help you find out which classes are a good fit for you.
- I will teach you about trades, degree plans, and financial aid.
- I will help you create a solid plan, so you have a clear direction.

- I have a proven track record as a college advisor and career coach. Together, we'll get you squared away quickly!
- Agriculture is my passion, and agricultural students are my favorite! You're an overlooked and underserved population, deserving undivided attention. I am here to serve you because your success is my happiness.

Somethings I believe:

- Agricultural students are our future.
- Agricultural professionals are our foundation.
- Agriculture is our fortune.

My book

What you can find in this book:

- All my college and career experiences have great lessons behind them.

- I poke a *lot* of fun at schoolin', farmin', and myself.
- Saddle up because I don't sugar coat much of anything. I tell it like it is, call it as I see it. No matter how humiliating of an experience I cultivated for myself.

Kick off your boots and sit a spell. *Don't Farm Naked* gives you a glimpse of my colorful college and career experiences. I've done it all wrong...and then right. These chapters are filled with must-have tips, stories, and lessons so you'll attain college success and will be happy in your career. My stories and lessons will help you get to the place you were meant to be: living a thriving life.

And if I can get to where I am meant to be, so can you.

“What’s the secret of success? Right decisions. How do you make right decisions? Experience. How do you get experience? Wrong decisions.”

– John Wayne

PART 1

TAKE THE RIGHT TRAIL

Figure out your schoolin' path

I Dunno

Why do you want to go to college? Is it because:

A. All your friends are going?

B. You don't know what else to do?

C. You want to get somebody to stop nagging you?

I'm here to tell you from experience: going to college for any of those reasons is not a great strategy.

After a year at Angelo State University (ASU), I decided to go to Southwest Texas State University (SWT, but now it's just Texas

State) because all my friends were there. Lunchtime was like being in our old high school cafeteria.

I enrolled, fell madly in love with my boyfriend, and then never attended any classes. Instead, I quit school, got a job, and made NO money (when you have to choose between rent or mac and cheese, you'll really see the light). Years later, I matured and decided I wanted an education after all. But by that time, I had seven Fs because I never attended, and I never withdrew. Here are some things about college I learned the hard way:

- It is much easier to recover from a W (withdraw) than an F.
- You have until the end of the semester to withdraw. Nobody told me!
- Trust your gut at the beginning of the semester. If the teacher or class doesn't feel right—HIGHTAIL IT OUTTA THERE

before the end of the first day. You want your full refund!

- Pay attention to your syllabus.
- Speak with your academic advisors.

And the crucial lesson: go to college for the right reasons.

"Of course, that's good advice," you're probably saying, "but what if I don't know what the right reasons are?"

Let me explain with a little story. In my first year of college, I put off a major English paper until the day before it was due. I went to the professor for help.

"You know it's due tomorrow, right? What do you want to write about?"

"I dunno."

Frustrated, she rattled off paper ideas like the War of 1812 and the tragedies of ancient Greece.

"Those don't really interest me," I said.

Rolling her eyes, Prof said: "Well then, what are you doing in college?"

"I dunno. My dad's paying for it, and he said I had to go."

Scary Statistics

From experience, I can tell you there is nothing more depressing or debilitating than living in a state of "I dunno." It's the feeling of having no options, direction, or believing you're stuck doing something you don't want to do.

However, when I first went to school, we didn't talk about these issues. And college can be a stressful transition. New roommates, classes, and teachers—oh my! I didn't know how to handle any of it. Nowadays, if you need it, there's a ton of support out there. I know most of you were raised to have thick skin, but if you start feeling down in the dumps, please reach out to your school's counseling center, professors, or academic advisors. It might lead to some serious consequences if you don't ask for help. Here are some scary statistics:

- Teen suicide is the 2nd leading cause of death among young adults aged 15-24.
- College-aged men are four to six times more likely to die by suicide than women.
- College-aged women are two to three times more likely to attempt suicide using nonlethal means than men.

These poor souls very likely had a history of depression, substance abuse, mental disabilities, physical disabilities, trauma, or abuse. They probably didn't understand themselves. They didn't know what was important to them nor have a solid direction that fit. In order to feel safe, they isolated themselves and got lonely. In the end, they didn't know where to turn or how to ask for help, perhaps because they hadn't found their herd.

Academic advising

More recently, I became an Academic Advisor for a junior college. Just know that some advisors, like me, take the time to understand your needs and encourage your hopes and dreams. Other advisors, however, herd you in and out like cattle, leaving you confused, your parents frustrated, and other advisors infuriated. It should be all about YOU from the get-go.

Personally, I help my students find the right path, make a plan, and feel good about their future. Here are some questions I ask my students:

1. Are you
    a. **extroverted** (outgoing, a people-person, etc.)?
    b. **introverted** (keep to yourself, not comfortable in big crowds, chatty people might drive you nuts, etc.)?
2. What degree do you plan to pursue?
3. What about that degree appeals to you?
4. What do you see yourself doing in your career?

Some students have a fairly clear idea about their answers to these questions. However, a lot more reply, "I dunno."

This would drive some parents, professors, and advisors nuts. However, it makes me giddy! Those students are my kinda students—especially ag students. For me, a confused ag student is like winning the lottery! I had been that student. I can understand that student and empathize. But most of all, I know I can help that student. Just like I know that I can help you.

You see, many don't understand that "I dunno" is short for:

I dunno... my options.

I dunno... my superpowers—my natural strengths, skills, and passions.

I dunno... the path that's gonna make me happy and make me money.

Whether you're fairly clear about your direction, or you're mostly thinking "I dunno," this little book will:

- Get you on the right path.
- Get you a plan.
- Get you feeling great about your future!

From the start, I want to help you get to where you're meant to be. You'll be:

- Living a life where you feel you have options.
- Living a life where you use your superpowers as much as humanly possible.
- Living a life that makes you happy and makes you money!

That way, you aren't sleeping on somebody's couch till you're 35.

And if I can get on the right path, get a plan, and feel great about my future, so can you.

What did the farmer tell his freeloader friend?

You're such a mooch.

Unlock Your Powers Unleash Your Potential

Remember those superpowers we mentioned in the last chapter? We all have them. They're our **strengths, skills, and passions**. Each of us is extraordinarily good at something. For instance:

- Some of you are laser-focused on science and math. You'd make you great veterinarians, ag engineers, or ag mechanics.
- Some of you are organized—your desks, backpacks, and horse trailers are neat and orderly. You'd be great farm operation managers, inventory managers, or financial officers at an ag bank.
- Some of you are go-getters, passionate, determined, and persuasive. You'd make great ag salespeople, teachers, recruiters, or extension agents.
- Some of you are resourceful—all you need is a little duct tape, some chewin' tobacky, and a Swiss Army knife, and you're out of a sticky situation. Being resourceful is a twofer: it's about solving problems and being creative.

> You'd make a great researcher, game warden (you gotta think like a poacher because those stinkers are creatively causing problems), disaster recovery specialist, or advertising specialist.

Now, this is by no means a complete list, and most of us are empowered by several gifts. My only question is: do you know *your* superpowers?

Even though you have these powers, you may not know what they are just yet. Sometimes it takes a process of elimination to figure them out. Take my brother, Bud, for instance.

My brother Bud's story

My younger brother, Bud, is an artistic, quick-witted, musical, creative genius who, at the age of three, drew an "oatmeal car" that ended up in every paper. However, just like they did with me, our parents drilled it in his head that he was going to college.

Not only did they tell him he was going to college, though, they also told him he was going to the University of Texas (UT), and if he even *thought* about going to Texas A&M, he was out of the will. I'm sure a lot of aggie parents have said the same thing to their kids about going to UT.

When Bud first went to UT, he was in the Longhorn band, played soccer, and majored in mechanical engineering. This had been a stable and secure fit for my dad, so Bud thought it would also be a good fit for him. As he got A's in band and soccer, he bombed the heavy math and science classes.

Eventually, Bud wound up on scholastic probation, was in summer school every summer, and was eventually "asked" to sit out a semester. Upon his return, he took a new class, where, as an assignment, he was supposed to be plotting out a highway exit ramp... but Bud thought it would look better in the shape of Bart Simpson.

His engineering professor, who did not appreciate his creative abilities, thought he would be better suited in a more artistic field. However, Bud heard our dad's voice in his head: "Art is not a real job!" So, he looked at the list of available majors on the back of the course catalog (long before the internet, of course) and circled business, architecture, and advertising. He didn't know what "business" meant, designing buildings sounded boring, so... advertising it was!

He changed his major, took a "study skills" class, and started a bunch of internships. As it turned out, advertising was a good balance of art and business. One of his internships turned into a job offer. And that job turned into a 27+ year career.

If Bud had focused on his superpowers when choosing his major from the start, rather than picking something because it was fit for our dad, he might have found his way to advertising and happiness sooner.

Today, Bud still uses a combination of his art and engineering superpowers, and he uses them for good, making incomparable monster masks and costumes. He also creates interview-worthy front yard displays for Halloween. In his creative process, Bud has become something of a *Halloween celebrity* in Texas.

How to find your superpowers

By now I'm sure you're dying to know how to find your superpowers. Let's sprout some ideas to help you:

List the things you enjoy doing; you're usually good at what you enjoy.

Get an outsider's opinion. Ask a friend, teacher, or someone you trust to describe you. You might be pleasantly surprised by what you learn about yourself.

List values that are important to you.

Take some personality and career tests.

Here are six freebies!

Want to get crackin' on those personality tests? These ones are free:

- 16 Personalities:

  https://www.16personalities.com/

- Crystal:

  https://www.crystalknows.com/disc-personality-test

- Test Color:

http://www.testcolor.com/personalitytest/personalitytest.php

- Human Metrics:

  http://www.humanmetrics.com/cgiwin/jtypes2.asp#questionnaire

- Texas Career Check:

  https://texascareercheck.com/ExploreCareer/InterestProfiler

- Ag Explorer:

  https://agexplorer.ffa.org/

Besides these free tests, there are also some not-so-free ones:

- CORE.
- StrenthsFinder 2.0.
- Big Five.
- HEXACO.
- DiSC.
- Myers-Briggs.

- PATH.

What do your results say about you? Talk them over with people you trust. Or better yet, invite them to take the test with you—they may not know their superpowers either. And if, after trying all of the above, you STILL don't know your superpowers, work with me. I'll grab my lasso of truth and pull 'em outta ya. Trust me; I have a *lot* of personal experience when it comes to finding superpowers.

My superpowers story

I always tell clients: don't go against your own grain! Long before I knew about personality and career tests, I went against my own grain *for years*.

I picked a boring, generic business degree plan even though my gut told me it wasn't a good fit. I should have picked communications. Instead of learning about managerial accounting—which I loathed and

NEVER used again—I could have been learning how to use my jabber jaw powers for good. I could've found my way to happiness sooner. But I heard my dad's voice in my head saying a business degree sounded credible, secure, and it would bring me better career opportunities.

Nope.

In my career, I also picked several boring desk jobs. You haven't met me yet, but I have ants-in-my-pants. I don't sit still *at all*. The only thing that saved my sanity in this kind of job was visiting with people in person or on the phone. Other than that, those poor career choices drove me crazy—stir-crazy and bat-crap-crazy. As a result, I got very depressed, drank like a fish, ate like a pig, and looked like a big ol' mad cow.

Since then, I've probably taken 100 of those personality and career tests. My results all say the same thing:

- She'll talk to a fence post.
- When she falls off a horse, she dusts herself off, and gets right back on.
- Preacher, teacher, shrink.

Now that I know myself pretty well, I also know life is short; too short to be doing things that go against my natural grain. When it came to getting my master's degree, I had learned my lesson: I wasn't about to settle for a boring, generic leadership degree. Instead, I went for that ag leadership degree from a top Texas school. I knew this provided the knowledge and skills I needed to focus on for the industry I wanted to be in. It's that choice that now allows me to do exactly what I love and love what I do—coaching and helping people like you.

Going with your natural grain instead of against it is the first step to building your personal brand, as it applies to both your schoolin' and career. You want to understand yourself. Because the better you know yourself, the better the choices you'll make *for* yourself. When you unlock your powers, you unleash your potential. And if you let that guide you, you'll become who you're meant to be—an *awesome* steward to others.

And if I can discover my superpowers, so can you.

"Courage is being scared to death... and saddling up anyway."

– John Wayne

Stand Out From The Herd

Branding goes back forever. A brand is created by using a unique, easy-to-read design, usually consisting of no more than two-to-three symbols or letters. Brands are intentional, strategic, distinct, and recognizable. European travelers brought the practice to the US where, to this day, ranchers used brands to take ownership, stake their claim, and send rustlers a strong message that says, "Steal my cattle, get a rope."

Today, companies use the same concept when branding a product or service.

Everything from the wrapper, message, colors, fonts, and placement in the store are intentional and strategic. You know you got it right when the brand's "personality" stirs emotion and puts the audience first. This builds trust, creates a positive connection, and crafts memorable experiences. When people become loyal to their favorite brands, it's hard to steer them away.

Personally, I'm a loyal fan of the show *Yellowstone*. (In fact, I'm an addict; I may need to find a Yellowstone Viewers Anonymous meeting soon.) In the show, it's the cowboys who're being branded. According to my man Kevin Costner, that hooked, rocking Y-shape of the brand represents a few things to those cowboys—belonging to something bigger than themselves, second chances, and trust. It's also a badge of honor, intentionally placed right across their bare chest. Only certain cowboys receive the brand, which makes those cowboys stand out—to everyone in that town and every loyal fan of the show, that brand is distinct and recognizable.

When it comes to college-and again later in your career-, your job is to create a personal brand that makes you stand out from the herd. Your brand is about you, but it ain't about you. Your wrapper—how you speak, dress, and behave—are extensions of your brand and should be intentional. But what is important is how you make *other people feel*.

When I went to visit with Dr. Tarpley at Tarleton State about the ag leadership master's program, I thought a head and came prepared. I thoroughly researched the program, looked through classes and descriptions to see if their classes were a good fit for me, and brought in my unofficial transcripts so he could see which of my classes transferred. In addition, I came armed with questions so we could have a productive visit. I wanted to make a good impression. I wanted to stand out from the herd and have him to see my brand.

What I aimed for was that when Dr. Tarpley thought of me or talked about me, he would feel I was a dedicated student who wasn't wasting his or my time and was serious about graduating. For the next few semesters, I kept up my brand. I built trusting friendships with fellow students and my profs, and we became loyal fans of each other. So much so, after graduation, some of my teachers were happy to give me excellent recommendations for teaching jobs.

Put those irons in the fire and brand yourself. Be strategic about your wrapper. Think about how you want to make other people feel. Be that hooked, rocking-Y. Be distinct and recognizable. Stand out from the herd.

Build self-awareness

Your brand starts with you. As such, it's important to understand what makes you tick—why you behave the way you do. It's knowing your strengths, skills, and passions (as we explored in the previous chapter!). It's also clarifying your beliefs and values, learning your likes and dislikes, and building better habits. You take all that information, and start doing what sets your heart on fire!

Let's sprout some ideas on how to help you build self-awareness:

- Picture your favorite person. This could be your mom, dad, grandparent, cousin, aunt, uncle, or new ag coach.
- List the qualities that make them your favorite person.

- Compare yourself to those qualities.
- Focus first on the good stuff. Then see which of your qualities need improvement and decide which you would like to strengthen. Then work on honing those.
- Stay aware of your behavior and whom you want to be and work toward personal improvement every day.

When you build self-awareness, you will soon start noticing the benefits:

- Improved connections and relationships with others.
- Increased creativity levels.
- Improved skills by recognizing what you do well and what you need to improve.

- Raised happiness by aligning your ideals with your actions.
- Increased understanding of how others perceive your behavior, making you a better leader.
- Strengthened personal and work relationships by managing emotions.

By now, I hope you've at least taken a personality test or two. If you haven't done this yet, good Lord, I handed you the information in a previous chapter! You didn't even have to look it up.

Now, some people just won't take these tests. My guess is they don't want to do it alone, or they're afraid of what they'll find out about themselves. Heaven forbid they find out they have imperfections like everybody else. However, in my experience, when you're willing to take a look at yourself, what you find is not *all* bad. Actually, it's pretty darn good.

Heck, you might even end up with a new favorite person!

I've done a lot of inside work, understanding and bettering myself, personally and professionally. From that work came great insight and clarity. My strengths include communication, empathy, and resourcefulness. I know my skills are public speaking and the ability to build immediate, trusting, and long-lasting connections. Finally, my passions are agriculture and animal health.

I put these parts together and, through a process of elimination, figured out what type of degree and types of careers could use my great qualities and would put me in industries I loved. Guess what? I got a dream job. I was doing what made me happy in an industry that made me happy. When you have that combo, there is nothing better.

Be a game changer

You may be like me and not have a competitive bone in your body. I was rarely a part of any kind of competition, and if I was, I'd wave off the others and say, "Ya'll go ahead." But throughout life, I've come to understand we compete for different things:

- Top percentage of your class.
- American Star Award.
- Grand Champion
- College acceptance.
- Grants and scholarships.
- Degree program acceptance.
- Jobs and promotions.
- A parent's love...

A little competition can be a good thing. Our drive kicks in and motivates us to work smarter and harder to get what we want. Also, competition teaches fair play, encouragement, and that there are wins and losses. It builds skills of perseverance, resilience, and tenacity. Since you're up against several challenges in life, you want to set yourself up with a competitive edge. This competition will also increase your courage, self-esteem, and self-efficacy. Just know that the true advantage really comes when we strive to be a game changer.

So, what is a game changer? This is someone who significantly alters the way things are done as a whole. Game-changers find ways to stand out from the herd by way of their personality. By mere virtue of their actions, these people can make changes that transform the landscape as a whole. Here are some benefits of being a game changer:

- You use your personality traits and attitude to spark change.

- You practice new and innovative plans, methods, and strategies.
- You revolutionize the way others think and do things.
- You build confidence and connections.
- You ride out any uncertainties.

How to be a game changer in college

TIP

Make your school tour a purposeful trip.

Don't worry about finding the library or cafeteria—if you've seen one, you've seen them all. Instead, stop by admissions to find out what you need to apply. Stop by Financial Aid and ask about grants and scholarships. Stop by the advising office and talk about your testing scores, interests, and classes. Take notes so you can get them what they need right away.

Next stop...Agriculture Department.

Before you go on tour, send the ag department an email asking to set up an appointment at a time that is convenient for *them*. Introduce yourself to those professors, shake their hands, and help them put your face to your name. Consider ag professors your most reliable source. They have department and industry knowledge and connections. Take the initiative and get them to like you. After all, people help those they like.

Respect your academic advisors and professors

If your financial aid hasn't kicked in, or if you waited until the last minute to register, or if you partied too hard the night before, or if you're a crabby person in general, DO NOT take it out on your academic advisor or professor. Speak to them as you would speak to your favorite person. They're there to help you. The better you treat them, the more they *want* to help you. And trust me, you want their help.

Don't waste other people's time.

Have your school ID ready anytime you come in or call the school. That's how they look you up and help solve your problems. Don't make them wait while you find your purse or your id in your purse. Don't make them wait for you to log into your computer and then scroll through a hundred emails to find it. Think ahead. Be ready.

Send emails that make them want to help you.

School departments are serving a ton of students at the same time. If you want them to respond to your emails more quickly, add important information they need. Don't send your fifty questions in one ginormous paragraph. Instead, number each one of them and space them out. In the subject line add your name and school ID number. At the end of your email add your name, school ID number, and phone number. Proofread your emails, out loud, before you send it so you know it makes sense. Make people *want* to help you.

TIP

Be prepared when you speak to an advisor.

This again includes bringing your school ID and driver's license. Also, advisors like it when you bring questions ready to ask. It shows you take initiative and are invested in your future. You've probably never done college before, so don't be afraid to ask questions.

TIP

Bring in support for support, but YOU do the talkin'.

Speaking from experience, there ain't nothin' more fun than dealing with overbearing, demanding parents or pillin' a cat. It's great to work with your support people to create advising questions so ensure you cover everything they want to know. But don't let them speak for you. Strap on your boots and be ready and willing to have an adult conversation with your advisor or professors. YOU do the talkin'.

Call advising with your own questions.

Advisors want to talk directly to you. DO NOT have your parents call them for you. In fact, advising can't even talk to them unless you have a FERPA on file. And for heaven's sake, don't let them call in and pretend they're you. Get up early and call advising when they open. Or call during your lunch hour. Or even look online to see when advising offices are open late. Or take off a few hours from work to see an advisor. Take responsibility and do it yourself.

If you don't like the academic advisor you're working with, get a new one.

It's *your* college experience. So, ensure the advisor is a good fit and that they're helping you move in the direction you want to go. If they're not helpful, it's your right and responsibility to get a new one.

Find the Academic Calendar and keep an eye on it!

At most colleges and universities:

- Spring registration starts mid- to late October.
- Summer and mini-mester registration starts around mid-March.
- Fall registration starts mid-March.

Know your registration dates. Know your payment due dates.

If you miss either of these, the school will drop you like a hot potato.

Don't dillydally and wait until the last minute to register.

YOU'LL GET THE LEFTOVERS... These are the leftover crappy classes at crappy times with crappy teachers who get 1-star reviews. These are the teachers who don't give a rat's behind if you pass or fail—they will try to weed you out.

Register yourself for classes.

Don't get your mom or dad to register for classes for you. Do it yourself. Be the first to register. Get the best classes, teachers, and schedules.

Advisors and teachers want to help you, and not just because they get paid to do it. All you have to do is behave in a way that makes doing their job easy, so they can help you because they *want* to, not because they *have* to. Stand out from the herd. Build their trust. Help them associate you with someone who has their manure together and wants to be successful. Take the initiative. Be prepared. Take responsibility. Be a game changer.

And if I can stand out from the herd, so can you.

A collie was talking about how hard he works on the farm where he lives.

A nearby sheep piped up "YOU don't work hard! All you do is boss us around."

"What did you say?", shouted the collie.

"You herd me", the sheep replied.

PART 2

COMING OUT OF THE CHUTE!

Navigate your way through school

Look It Up

Long before Google was a thing, we had a physical dictionary, thesaurus, and encyclopedia. Whenever I didn't know something, though, I'd bypass all those resources and go straight to my mom for the answer. She was my Google. Still, most of the time, she didn't give me the answers freely, even when she knew them. Oh no. Her favorite reply was: "Look it up."

I remember how I'd rolled my eyes at her. If I wanted to walk across the room, dust off the dictionary, and find the answer myself, wouldn't I have done so already? I honestly think that Google was created by someone who got tired of their mom saying, "Look It Up."

Now, there were two potential reasons why my mom told me to "Look it Up."

1. She didn't know the answer either (and wasn't willing to fess it up).

OR

2. She wanted me to learn how to be resourceful and find the information *for myself.*

When you go to college, academic advisors will be there to guide you through your college career every step of the way. If you have questions and can't find the answers yourself, advisors are an excellent resource.

But at least *try* to find the answer yourself first.

After all, you don't have to walk across the room and pick up a book anymore. All you have to do is pick up a device—any device—and...

LOOK IT UP!

Advisors aren't trying to hurt anybody's feelings. But just so you know, when you ask a question, and your advisor says:

"When I Googled it..."

OR

"When I went to the website..."

OR

"Let me show you where you can find it..."

These are just polite ways of saying: "Start using your head. That's the lump that's three feet above your ass."

Your school knows you have tons of questions. This is why your school's website is FILLED with useful information.

Frequently Asked Questions (FAQs)

Most of the answers you need are here.

Search Field

Type in keywords. It works just like Google.

Student Profile

This is where you can watch videos, read articles, or upload everything.

Academic Calendar

For registration dates and the start and end dates for classes.

Degrees or Certificates

This will show the programs and the degrees. When you click on *Degrees*, you can find department contact info (hint, hint).

Admissions

Send everything here: official transcripts, diplomas, shot records, etc.

Financial Aid

FAFSA: grants, scholarships, student loans, veterans (GI bill, Hazelwood), etc. More to come later on this info.

Academic advisors are the gurus of the school. However, when there are a billion other students who also aren't looking things up, you'll probably end up on hold; *forever*. Save yourself time and frustration—learn to find the information *yourself*. It gives you a sense of accomplishment and boosts your self-esteem. It empowers you.

It's not your first rodeo, so look it up. If you can't find the answers, then, and only then...

And if I can learn to *look it up*, so can you.

Tricks Of The Trade, School And Otherwise

I believe what parents want most is for their kiddos to live an abundant life, to be self-sufficient, and not to suffer through hard times. As such, they want you to make choices that set you up, as much as possible, for a future-proof life—meaning that employers continue to see your knowledge and skills as valuable. If you have essential skills, even during tough times, you'll more than likely have a job, can put gas in your truck, and can feed yourself and your family something other than mac and cheese.

Education (be it formal or something else) gives you knowledge and skills that keep you self-sufficient, even when times get tough.

True story. For instance, when Covid hit:

- 98% of my *out-of-work* clients had an associate's, bachelor's, or master's degree with work experience. Depending on the industry and their pay scale, clients included account executives, marketing managers, lawyers, and engineers.
- 2% of my *out-of-work* clients had solid work experience but typically no formal degree. Depending on the industry, these clients included executive assistants, customer service reps, and help desk reps.
- 0% of my clients who had essential skills or a trade under their belt were out of work. They kept this country running.

> We worked together on learning resume writing, interviewing skills, and how to leverage their superpowers with employers so they could negotiate better pay, better benefits, and better careers.

Fortunately, this last group had set themselves up to be future-proof with in-demand skills. They weren't afraid to get their hands dirty, and to employers and the rest of the world, they were essential. Included in this group were electricians, plumbers, surveyors, medical assistants, landscapers, HVAC, public safety workers, first responders, police officers, utility line workers, truck drivers, diesel mechanics, welders, respiratory therapists, legal assistants, registered nurses, IT technicians, auto mechanics, and veterinary technicians.

As I mentioned before, I quit SWT after I never went to any classes. I figured I just wasn't "college material" yet and decided to go another direction.

I can tell you my parents weren't happy when I told them I was quitting school, but it allowed me to test out some things I felt I would enjoy. I worked as a waitress, was happy, and turned that into a cash cow career--*Ka-ching*! During that time, I spent a few months going to a trade school. I learned a new skill and got a certification doing something I thought I would like—be a nail technician. That certification fit me well, nail techs in-high-demand, and it sustained me for years. Especially when I went back to college for reals.

Of course, the school tried to convince me to go through the entire cosmetology program, but I told them I didn't even like messing with my own hair; why would I want to mess with somebody else's?

Not everyone's college material, and there's no shame in that. Just because I have my master's degree doesn't mean I am "college material," either. I don't like reading or researching or any kind or math or science. I also have ADHD tendencies, so sitting through uninteresting classes was very tough for me. In fact, I'm not quite sure how I ever graduated.

I believe if someone had warned me that the first two years of college were basically 13th and 14th grade, I might've gone in another direction sooner instead of wasting my parents' hard-earned money. As it turned out, trade school was a much better fit for me because it was short, hands-on, and fit well with my desire to help others feel good about themselves.

So, I agree with your parents—I, too, want you to live an abundant life. I want you to be self-sufficient and thrive during tough times. Therefore, I'm a big fan of education. But you have to make sure it's the *right kind* of education for you. Don't waste your time or your parents' money if you don't have a desire to repeat 42 hours of core classes that you just got done doing in high school.

To give you an idea of what those core classes look like, in Texas, these are: two English, two History, two Government, two Math, two Core component, two Social and Behavioral Science, one Creative Arts, and two Science (TAMU wants three or more, depending on your major) classes.

If you're reading that and think, "that's not for me," then let's instead figure out your superpowers. Let's find you a trade that will make you happy and make you money.

With a trade:

- You will be in high demand and probably more future-proof than many others.

- You will be self-sufficient when times get tough. And if, like me, you decide to go back and get your degree, you'll have a skill you can use on the side, anytime.
- You will make more money faster because you will enjoy what you do, and you will turn it into a cash cow career!

Here are some tricks of the trade, school and otherwise, you need to be aware of.

School is a business.

A school's main goal is to give you a good education, but they are still a business and are therefore out to make money. It doesn't matter if they are private, Christian, public, or the Xavier Institute. All schools depend on their admission and retainment numbers to stay in business and to get funding.

TIP

Don't take the bait.

Although having an education is always better than not having one, sometimes schools will use "means to an end" practices to get you in. Clickbait is a questionable advertising method, usually in an ad, image, or headline, that is designed to *bait* readers into clicking on a hyperlink. However, once you click on the link, your information is then captured and then sold as a "lead" to anybody and their dog.

TIP

Don't get trigger happy.

When you are on Instagram, Tik Tok, Snapchat, or any other site, don't be so quick to click on everything that looks interesting. Private and public schools deliberately post ads, images, or headlines that grab your attention. These are considered "means to an end" practices and can be shady. If it ends up being clickbait, the company creating it will sell your information to about a hundred schools who then proceed to blow up your phone, email, or snail-mail box in order to get you to attend their schools.

Not every job posting is for reals.

When you look for a job on *any* job board such as Indeed, Zip Recruiter, or Simply Hired, schools will post an ad, image, or headline, and word it like a job posting. When you click on that link, the job boards will sell your contact information and list you to about a hundred schools or other job boards. Some of my career coaching clients who got a little trigger happy-clicked on something that they shouldn't have-and received an endless amount of phone calls and up to SIXTY emails a day they never signed up for.

Don't just delete!

When *you* chose to click on that ad, you give both the schools and job boards the legal right to stalk you until the end of time. The admission rep's job is to get you to come in and sign up for their school. The person working at the job board is trying to continue to sell your information. If an admissions rep or job board worker doesn't do their job, they will lose their job. This is why they often act like a used car salesmen using ruthless tactics to get what they want.

If you end up on somebody's list, here is what you can do:

- Don't just delete their email or phone calls. This gives them permission to keep contacting you. Instead, unsubscribe or block them.
- Tell them verbally or in writing-to have record of what they said, "Do not contact me." or "Take me off your list".

Do research on tuition cost versus salary outcome for the trade or degree you choose.

Some trades and degrees are so generic that earning them will not pay off after you graduate. I hear students say they want a degree in arts, philosophy, or general studies, or train as a CNA, manufacturing technician, or carpenter because that's what interests them. I'm all for doing something that makes you happy, but before you shell out money for your trade or degree, you better ask yourself, what the heck am I gonna do with that?

I want you to be happy, but consider what you will do with that degree. Earning a living is just as necessary as following your passions. Make sure that you create a plan that lets you earn a living while doing something you love. There is a big difference if you're eating mac and cheese every day because you want to and not because you *have* to.

Remember, schools are a business and many schools have been shut down for their shady business practices. For example, at one school, a culinary degree ended up costing students $90,000. That school assured students they would make twice that salary when they finished. However, when they got out into the real world, they found out that they needed more experience and education to make the kind of money there were promised. The only job they could find was being a short-order cook at local grill, making $10.00 per hour. That school was finally shut down for gauging students.

- Know what the heck you are gonna do with that trade or degree. Talk to people in that field before you sign up.
- Know whether or not the money you shell out for the program is it's going to pay off.

Forget the Associate's Degree!

Except for the school offering one, nobody cares about you having an associate's degree! I've looked at tons of job postings. Employers will want to see a bachelor's degree, sometimes even a master's degree for the job college graduates are looking for. With the convenience of getting a degree online, *many* employers will pick a candidate who has a bachelor's degree over someone who only has an associate's degree.

Exception to school rule.

If you think I'm full of bull and *really* want to go for an associate's only degree, don't pick something just to have something. Do your research and pick *something* in high-demand. If you want to go to school, then get hired, and paid, get something that will make you *essential*.

- Check out O NET Online to see whether or not your interest has a bright outlook. https://www.onetonline.org/.

Make educated decisions about the right program for you. Here are a few common differences between a private trade school and public-school technical programs.

Admission requirements

Private:

- Students ARE NOT required to complete ACT/TSI/SAT testing for certain certifications.

- Students ARE NOT required to take core/basic classes for certifications.

Public:

- Except for certain certifications, students ARE required to complete ACT/TSI/SAT testing for most programs.
- Except for certain certifications, students ARE required to take core/basic classes.

Time frames

Private:

- Start dates: Every 3-5 weeks.
- Day classes take about 1 year with regular attendance and passing grades.

- Night classes take about 15 months with regular attendance and passing grades.
- Hands-on classes are readily available.
- Students dive into hands-on learning almost immediately.
- Programs may take less time to complete than public.

Public:

- Every spring or fall semester, with some available summer classes.
- Classes for popular programs fill up quickly. Don't dillydally around.
- Technical program classes are not always available, so public schools often *encourage* students to take core classes while you wait to get in to your technical classes.

- If you can't get into the technical classes, the public program may take longer to complete than the private.

TIP

School cost

Private:

- May initially seem more expensive, but if you are not required to take core classes while you wait to get into the public program, it may cost less.
- Offer in-house scholarships to off-set the cost.
- Students are usually required to pay *something* monthly while in school to offset future debt.
- Paying *something* as you go helps students take classes more seriously and off-sets future debt.

- Financial Aid department tries to make monthly payments affordable.
- Typically, no housing is offered.

Public:

- Might cost more if you take core classes while you wait to get into the technical program.
- Offer in-house scholarships to off-set the cost.
- Don't require students to pay monthly.
- The entire school debt is left until you finish, making you see financial aid as "free money."
- Financial aid (govt funded) only like paying for traditional degrees (AA, AS, AAS), and often won't pay for certifications.

- Housing is typically offered.

TIP

School support

Private:

- Fewer students, more support.
- Admission reps are out to make a sale and will stalk you to get you to sign up.
- After you get settled in to school, some career services departments will connect you with employers so you can work while you go to school.
- As you get closer to graduation, the career services department usually helps students find internships.
- If you behave at the internship, it can lead to permanent placement.

- They are a business with numbers to meet (which is the reason for all of the above).

Public:

- There ain't enough money or manpower to support the number of students.
- Public school departments will usually be more laid back.
- They are a business with numbers to meet, but without the right amount of manpower, they often herd you in and out like cattle.
- Career services resources are available, but you'll find your own job after graduation. Impress your professors and created a strong network along the way to CYA.

School ownership

- Most trades schools are privately owned.
- Texas State Technical College (TSTC) is the only state-owned public trade school in Texas. This means TSTC offers housing and FA for programs, including many certifications.

Look through trade school or junior college technical programs and certifications to see what might interest you.

Talk to those departments to see what a career and salary in this field look like.

Talk to people in that role and ask if you can do a ride-along. That way,

you can see what a day in the office really looks like.

See the list salaries for Ag-related essential services on the next page.

It's been 30+ years since I became a nail tech, and I still have my original case. Although my trade kept me self-sufficient for years, these days, that case is filled with polish from hot pink to sparkly, neon purple. Whenever I paint the little neighborhood girl's fingers and toes in "Unicorn Colors," it brings both of us happiness. It makes me thankful for learning those skills long ago. I *highly* encourage you to get edumacated, but make sure it's the right kind for you.

Be happy. Be essential. Be self-sufficient.

And if I can make choices making me future-proof, so can you.

Ag-Related Essential Services Pay Scale

Could be MUCH higher depending on education, certifications, licenses and experience

Carpenter	$16 - $52 per hour
Commercial Driver (CDL)	$26 – $38 per hour
Crane Operator	$28 - $46 per hour
Diesel Mechanic	$19 - $48 per hour
Facilities Maintenance	$20 - $29 per hour
Plumber	$28 - $45 per hour
Electrician	$19 - $40 per hour
Heavy Equipment Mechanic	$25 - $37 per hour
Heavy Equipment Operator	$21 - $48 per hour
HVAC Tech	$24 - $40 per hour
IT Tech/Helpdesk	$17 - $26 per hour
Landscaper	$15 - $35 per hour
Septic Tank Tech	$20 - $30 per hour
Survey Tech	$20 - $28 per hour

Veterinary Tech	$17 – $20 per hour
Water Well Digger	$25 - $40 per hour
Wind Turbine	$27 - $38 per hour
Welder	$22 - $30 per hour

Why did the farmer plow her field with a steamroller?

Because she wanted to grow mashed potatoes!

What is a happy farmer's favorite candy?

A Jolly Rancher

Getting To The Root Of School

Most bachelor degrees require about 120 credit hours to graduate, which includes your core classes. However, some degrees require that you take some extra science or math classes. So, make sure you know how many hours of science and math are required for your major at the university you plan to attend.

Quick lesson: course numbers

One way to see how many credit hours you get for a class and what level the class is, is by looking at the course numbers. Here's how that works:

- **First Number**: The level of the class.
- **Second Number**: The amount of credit hours you get for taking the class. This applies toward the total amount of required degree hours.
- **Third/Fourth Numbers:** These refer to the subject's sequence of classes.

Still confused? Here are some examples:

- 1119 = A level 1 class, with 1 credit hour. This might be a lab class only, Kinesiology (KINE), or PE-type class.
- 1201 = A level 1 class with 2 credit hours.
- 1301 = A level 1 class with 3 credit hours.

- 1401 = A level 1 class with 4 credit hours. This means there is a class (lecture) + a lab.
- 2401 = An upper-level class with 4 credit hours, usually the sequel to the 1401 class.
- 2402 = An upper-level class with 4 credit hours, and the sequel to the first upper-level class (2401).

Texas schools using the three-course numbering scheme is as follows:

- 100–199 - mainly open to freshmen
- 200–299 - mainly open to sophomores
- 300–399 - mainly open to juniors
- 400–499 - mainly open to seniors.

Lettuce talk celery majors

STEM classes --Science, Technology, Engineering, Math, and celery

If you have a STEM major, you may be required, to take math classes in this sequence:

1. MATH 1414 (College Algebra)
2. MATH 1316 (Trig)
3. MATH 2412 (Precalculus)
4. MATH 2413 (Calc Differentials)
5. MATH 2414 (Calc Integrals)

Additionally, about 3 more upper-level, hard-core, classes.

You better *love* math or at least have done well in high school precalculus (pre-calc) classes. If you haven't taken or done well in pre-calc, you may want to rethink your major. If you're still certain you want to major in Computer Science, however (maybe because you like playing video games), know that you also need to have taken and done well in College Algebra and Trig in high school. Typically, pre-calc builds off these classes and gives you a foundation for college-level pre-calc classes.

Instructors aren't going to teach you the basics in a college math class; they will expect you to understand these already, so they can move forward with upper-level material.

You also better *love* science and make high grades. You won't have a lot of choice in class choices because the school will tell you which sciences they want you to take. So, make sure you take the classes they will accept, or you could end up wasting your time and money.

When you apply for STEM programs at universities, they look *very closely* at your math and science grades. Keep your grade super high in those subjects so you can get into the best programs.

Go over the transfer guide with an advisor.

Why are chemists great for solving problems?

They have all the solutions.

How does a veterinarian tally up a patient's bill?

They use a cow-culator!

Name the favorite dog of Charles Darwin

A lab.

Who on earth make the best married couples?

Carbon and Hydrogen

They bond well together since their first meeting

Once there was a hen who counted her own eggs.

She was a mathemachicken!

What do you call a computer superhero?

A screen saver!

Did you hear about the engineer who invented the escalator?

They were mechanically inclined.

Lettuce talks tater majors

STEM-less classes—Communications, Leadership, Business, and taters

You're my people. Personally, the only time I ever enjoyed math was in the 3rd grade, when I had to learn my multiplication/times tables. I remember I wanted to impress my dad, an engineer who was brilliant in math. Luckily, memorization was my strong suit so I used this to my advantage. And you can bet my dad was impressed with how fast I memorized everything. To this day, I can still tell you what 9 x 12 is without using a calculator or my fingers and toes.

However, when it came to choosing my major, I had a strategy: pick the one with the least amount of math and science. I was doing just fine this way until somebody slipped statistics into the mix—I thought I was gonna DIE.

So, if math and science are not your strong suits, or you loathe them *and* celery like me, here are some tips to keep in mind:

TIP

Many universities will NOT count college algebra as a core math class. They will require you to take MATH 1324 (Business Math) and MATH 1325 (Business Calculus).

TIP

If you take dual credit math classes in high school, don't waste your time taking college algebra. Instead, take MATH 1324 and MATH 1325.

TIP

Depending on your major and university, they might accept MATH 1332 (Contemporary Mathematics). Although the number implies it is higher than MATH 1324, some students consider it easier—especially easier than Business Calculus. Of course, the level of ease will depend on the material, your math skills, the instructor's teaching style, and the effort you put in.

Most STEM-less majors allow you to choose your own science classes. However, not any ol' science class will do. The university where you plan to transfer has a list to choose from. Choose your classes wisely.

If you have to sit through a science, you might as well pick classes you find interesting, have done well in, or believe are a good fit for your career. Just make sure the class is still on the list.

If it were me, I'd probably choose the lowest level of science that would still be accepted at a university. There's no way I would suffer through another chemistry class for an ag communications, leadership, or business degree.

Let's sprout some science ideas for my non-STEM peeps:

- Horticulture (AGRI 1315 and 1115 (lab)): it is important for EVERYONE to know how to grow their own food. I needed this class because I kill cactus. Besides, ag people have to show regular people that food doesn't come from the grocery store.

- Animal Science (AGRI 1319 and 1119 (lab)): this might be a repeat from high school, which means it might be easier the second go-round. This was the only science class I took, and we were castrating piglets in week one! Never had to do that in algebra. It turned out to be the hardest but most memorable class I ever took.

Anthropology (ANTH 2301 or 2401 (lab)): it's good to study societies and cultures. If you want to communicate with other people on their level, you need to understand people and where they came from.

Geography (GEOG 1301): this is similar to Horticulture. It's the study of the earth, land use, and maps.

Geology (GEOL 1301): this is the study of fossils and rocks. I actually got nostalgic so I whipped out my old rock tumbler to make some tacky keyrings—which I loved!

Oceanography (GEOL 1445): this is basically the study of the ocean. I must be part golden retriever because I love the water—especially the beach. People who learn to understand the ocean and coast often treat it better.

Remember: when picking classes, whatever you do, DON'T GO ROGUE! Speak to an academic advisor and go over the transfer guide, so you know what classes are required for your major and university. Set yourself up for success. Choose your majors and classes wisely, but also make your college experience enjoyable.

If I can choose classes that are right for me, so can you.

I'm 25% funny and 85% bad in math.

Where does a farmer get his medicine from?

The farm-acist

In my Horticulture class we were supposed to grow cucumbers, but I didn't pay attention to the lessons.

You could say I was in a pickle

If you are looking for a job, get trained in horticulture.

It's a growing industry.

I can speak in public and in front of large audiences.

What's your superpower?

“Agriculture engenders good sense, and a good sense of an excellent kind.”

- Joseph Joubert

Good Sense About Your Cents

College is very expensive these days. If you can swing it, the best thing you can do is pay for college out of pocket as you go. The upside is, you don't owe anybody when the class is over, you have control over how many and what classes you want to take, and you can take classes whenever you want to take them. The downside is, it can put a dent in your pocketbook and slow down the process of finishing up.

I am by no means an advocate for the "buy now, pay later" mentality. Once you get into that debt cycle, it's very hard to get out. I've spent most of my early adult life owing other people. Owing other people sucks the energy, courage, choices, and life right out of you. Nevertheless, when it comes to investing in yourself, receiving financial aid is an alternative route. You'll be finished with your degree sooner, and this way won't strap you for cash during your education.

However, remember this: he who holds the purse strings holds the power, and nothing in life is free. Somehow, some way, somebody has to pay for it.

How financial aid works

Financial Aid (FA) is money from the federal government to help you pay for tuition, books, housing, and food while you go to school. The amount you qualify for is based on how much money you or your parents make.

When you apply for financial aid, you need to fill out a FAFSA application online at www.fafsa.ed.gov.The process will tell you if you qualify and what type of financial aid you qualify for.

Here are some tips about FA funds:

Student loan: this is money you pay back with interest at a lower rate. Repayments begin about six months after you graduate and no, payments are not considered "optional".

Pell Grants: these are commonly referred to as "grants." With these, you get money you don't have to pay back.

Scholarship: this is money you don't have to pay back.

Other funds, separate from federal funds, are:

Education loans: these come from credit unions, banks, or even loan sharks. They're considered private loans and come at a higher interest rate. Payments, again, are not optional.

In-house scholarships: these come from the school you attend. As with regular scholarships, you don't have to pay back the money you receive.

Outside grants and scholarships: search for these on the internet. There is a ton of money out there that you don't have to pay back. You just have to look it up.

When you apply for FA, and if you're under the age of 24, you will need to use your parents' latest tax return to show proof of income. If, for whatever reason, your parents don't have a tax

> return, you won't be able to apply for FA until either they get one or you turn 24 and can show proof of income.

The reason for this is the following: long, long ago, somebody who was out of their right mind began awarding young adults more money than they knew what to do with each semester for school. Many students saw this as "free money," bailed on school, and flew the coop. There was no consequence. What were they gonna do, come take back the classes you took? No collateral made it hard for the government to collect.Then someone in their right mind figured out that most young adult students, probably aren't that financially savvy or responsible. No kidding? The government decided to hold parents responsible for paying back the money because parents usually have some sort of collateral to back up the loan.

> It's important to know this: if you or your parents take on a student loan,

it's not "free money." Although you "buy now, pay later", this is *loaned* money only so you can have a better future. Come time, though, the government will come a callin' because they want their money back. They usually get it, one way or another—automatically from paychecks, tax refunds, or from a lien on your home. If they have to, they'll try to take it out of your hide.

Even if you're awarded grants or scholarships, it may seem like free money, but there's still a price to pay. You will be required to follow the government's orders. They will dictate how many classes you must take each semester. If you screw around, fail a class, and your GPA tanks, they will cut you off and it will be tough to get back on. If you don't withdraw from classes before the withdraw deadline, and you fail the class, they will cut you off. If later on, you find out that the class you failed is required to graduate, they can refuse to repay for that class. This means you'll have to retake and repay for it out of your own pocket. Kinda defeats the purpose of getting financial aid.

Credit Cards

Beware!

Credit card companies stalk students. Earlier in the trade school tips section, I warned you about "clickbait" methods. Just know the same shady ads, images, and headlines are created for college money to *bait* you into clicking on the hyperlink. When you click on that link, credit card companies buy your contact information off some list. Then, they proceed to blow up your phone, email, or snail-mail box. They know you want school money. And they know how good it feels to be "approved." So, they'll make you an offer you can't refuse.

How credit card companies work

First, credit card companies “invite” you to apply for a card with an introductory offer that has no interest for the first 18 months. Then, they highly encourage you to use it “just for emergencies” or on things you need, so you get “3% cashback on your purchases.” This feels like free money. It’s not.

Credit card companies play into your emotions and insecurities. Once you’re “approved,” they send you a shiny, new, personalized card with a picture of your school, horse, or dog on it.

If you’re financially savvy and responsible, this might work. However, they know most young adults and many parents aren’t. This is what they’re hoping for! Credit card companies usually make back their money

100 times over in interest, late fees, non-payments, and lawsuits—yes, they can sue you and will win when you can't pay up.

They know that some of the charges will be for school necessities. However, they also know most charges will end up being for things like a spring break trip, concert tickets, or beer. Credit card companies are betting that at some point, for instance, when you go out to dinner with your friends, you like the feeling of being approved, so you're gonna whip out that card and pick up the check. That card makes it seem like it's free money.

What's more, if you max out your credit card, those companies will do you a favor and increase your limit! How neighborly of them. They look at you as a cash crop because they know if you maxed out once, you'll probably max out again.

However, since your parents are probably the cosigners on the card, they'll get the astronomical bill—not once, but twice—which puts a dent in their paycheck. Next thing you know, you get a call from one of your parents, addressing you by your first *and* middle name—that's never good. Then, they take away your allowance, threaten your life, and advise you to cut up the "damn card" or else.

As I said before: he who holds the purse strings holds the power. Nothing in life is actually free. Somehow, someway, somebody will pay for it.

Financial tips for your college experience

Feeling unsure and overwhelmed by how on earth you're going to pay for college? Here are some tips to help you along:

Start out at a two-year community college and then transfer to a four-year university. The average cost of a two-year school is $700 per class. The average cost of a public four-year university class is $1,100. Truth be told, most young adults aren't crystal clear about their direction yet, much less if they want to be in college. Do yourself and your parents a favor and, while you're figuring things out, start out at a community college. It will be cheaper. In addition, the transition from high school to college will be smoother. And as a bonus: you're still gonna get a good education because many instructors who teach at four-year universities also teach at community colleges.

I am a HUGE fan of veterinarians and applaud anyone wanting to get into that line of work. However, if you're tossing around the idea of going to veterinary school, you do need to be prepared. So, while I'm not trying to spread manure all over your dream, here's what you need to know:

- The competition to get into vet school is crazy fierce. Your best chance of getting accepted is to maintain a 4.0 GPA. You better have straight A's for your math and science classes, especially.
- According to the American Association of Veterinary Medicine Colleges (AAVMC), the 2021 average cost for a 4-year veterinary program runs over $200,000. That's JUST for vet school. This does *not* include the time or money it took you to get through your bachelor's degree. Once you get past residency, where you'll make $38,000 per year for THREE years, the veterinarian's average salary is $70,000.

Make an educated decision. You don't have to know if you are going or not in a red-hot minute, so keep your GPA up just in case. And, after you do the math, if you still feel it's worth the time, money, and effort, start looking around for grants and scholarships NOW. When the time comes to apply, you want to already be familiar with the money that's offered, along with application deadlines. That way, you can be one of the first to get that free money.

TIP

PS: although scholarships are out there for both small and large animal medicine, there's more money available for the large animal option. This is because many students are focused on small animal medicine.

TIP

Stay in state. This means that you go to school in the state where you live. Let's say your significant other, who has money coming out of their ears, decides they want to go to school out of state. You believe you can't live without them, so you decide to follow. Unless your partner is willing to pay for you, too, stay put in state. If you go out of state, you or your parents will pay out-of-state tuition for at least a year while you set up residency. The cost is typically more than three times as much and not worth it. Especially if you are on FA, because you are the one responsible for paying that money back.

Go to a public school, at least for the first two years. Unless you're Richie Rich or have grants and scholarship money coming out the *wazoo*, tuition costs a bazillion times more if you go to a private school. And for goodness' sake, don't go to an out-of-state private school, ever. That's about as expensive as it gets.

If you have financial aid questions, call financial aid! Don't call admissions, advising, or the cafeteria. With the exception of the cafeteria, other departments only have high-level access to a student's financial aid records. Since FA is considered the money guru, the people here will be the best point of contact for discussing your financial aid in detail. So, save yourself some trouble and get the information straight from the horse's mouth.

Know if your parents are taking on financial aid, or if they're paying out of pocket. This might help you be more responsible about finances.

Be clear about what kind of financial aid you're receiving: grants, scholarships, Hazelwood, VA benefits, or student loans.

TIP

Ask FA for a breakdown of how much you're receiving in grants, scholarships, or student loans. Do what you can to only take on grants and scholarships.

TIP

Ask FA how many hours you must take each semester per your financial aid distribution. I had kiddos in my office who dillydallied around and didn't apply for FA until very late in the semester. However, since they were getting "free money," they were told they would be required to take four classes in one eight-week semester to keep their FA! Instead of waiting for a new semester to start or paying out of pocket, they signed up for classes, got overwhelmed, bailed on school, and left their parents to pay back their school loans.

TIP

Ask the school or search on their website for the breakdown of costs for your *actual* needs each semester, like tuition, books, housing, and food.

FA may approve you for more money than you really need. Know that you don't have to accept *all* the money they're willing to lend you. Break down the true costs of your actual needs, like tuition, books, housing, and food. Only accept money that will cover these expenses.

If you do accept more money than you need, you'll probably use it for other things. But trust me, you don't want to be paying 6% interest for the next 20 years on money you used to buy beer. That's a very expensive 12-pack.

Be willing to work, at least part-time. This can be your play money or a substitute for student loans. It will offset expenses or relieve some of the financial pressure and panic attacks those loans WILL cause later.

Join the military. You'll travel, learn self-discipline, and get a new perspective on life. Plus, they'll pay for almost any kind of school you want to attend. They're kind of like Richie Rich—they'll pay for in-state, private schools or even out of state, private school. In Texas, if you use up all the money from your GI bill, the Hazelwood Act kicks in to give you more school money.

If you aren't sure that you're college material, then don't put *anybody* in that financial position. Have an adult conversation with them about it. Tell them you'd like to work to get crystal clear about your direction before you move forward with school. If you work with me, I can help with that. Let's figure out your superpowers and get you some skills that will make you happy, make you money, and make you more future-proof.

Later on in life, when I went back to college for reals, I took a couple of classes per semester at a community college, while working a couple of jobs, and paid for school out of pocket. The company I was working for provided complete education reimbursement for the last two years of my bachelor's degree. So, it is possible to work your way through college without going into debt.

Except that, somehow, I ended up with close to $30,000 in debt?

Know why? I took out more student loans than I needed and got a credit card! I swiped that shiny thing until my dog's picture wore off. I told myself I would pay it off later, once I got a better-paying job—which, of course, I'd get after receiving that degree.

Those loans and the card felt like free money to me. So, even with that degree, I paid the price. It took me 20 years to pay back $30,000. And it sucked—the energy, courage, choices, and life right out of me.

Yep, I did things all wrong...and then right when I got my master's degree.

However, it was a learning experience. That's why I now share it with you, so you don't have to make the same mistakes I did. Take the tips you like and leave the rest. But whatever you do, do the opposite of what I did.

Go for your dreams, get the education you need, but be smart about taking on debt. Search and apply for money you don't have to pay back. Be willing to work while you're in school. Join the military and then have them foot the bill. Don't take on more money than you need or can handle paying back. Get clear about your direction. Remember, nothing in life is actually free, so use good sense about your cents-get savvy about finances and stay responsible.

And if I can learn to pay my way through college with minimum debt, so can you.

Why did the farmer fund his schoolin' through a bank?

Because borrowing money from a loan shark could cost him an arm and a leg.

Find Your Herd

Choosing good stock is a selective process used as a tool for improving highly desirable characteristics to be passed down for generations. Sought-after qualities include mental and physical health, nurturing abilities, adaptability, temperament, and behavior. Culling certain animals without those traits is necessary to prevent harm, loss, and other stressful problems. Having a herd with excellent characteristics will improve financial stability and boost living conditions for the owner.

Earlier, we talked about the dark side of college and some of those unfortunate statistics. Since college can be a stressful time, it's important to be just as selective as the farmer choosing his stock when it comes to choosing your herd of friends.

You might already have a strong support system and know your basic direction when it comes to college. However, new roommates, organizing and balancing responsibilities, and homesickness can be big adjustments. And all this overwhelm can bring on intense anxiety. When I was in that space, I had panic attacks at least three times a day.

Like many of you, I've always made friends easily because I'm pretty neighborly. When I was a young adult, the more friends I had, the better; it was about quantity over quality. However, when it came to picking friends, my picker was broke. Meaning I had friends in *low places*. Still, I considered those friends to be true and was certain I'd found my herd.

Later, I realized most of those friendships were superficial. Sadly, the more superficial the friendship, the lonelier I felt. The lonelier I felt, the more I became discouraged to move forward on my own.

Finding the right herd is important. For example, during a time when I was considered a Large Marge, me and my good friend, Dardi, decided to be exercise buddies. We started walking every day after work. I bought new walking shoes, shorts, and a bunch of healthy snacks. I was ready. I was pumped. I was feeling good!

Then on week two, I started making excuses. I whined, "I've got critters to feed. I don't have time. I've had a hard day..."

Dardi and I worked in the same attorney's office, about five feet apart, so we basically had the same day. However, she let me off the hook and said, "It's okay, we'll walk tomorrow."

Then tomorrow came. And when I tried to pull the trick again, Dardi said, "That's okay. But have you seen your butt?"

She didn't tell me what I wanted to hear, but she told me what I needed to hear. If it weren't for my friend, I would have gone home and gone to bed. Instead, I got back on the walking wagon and dropped 20 pounds. This is why Dardi is part of my herd.

Want to know how you can find your herd? Here are some highly desirable characteristics to consider for your herd:

- Friends who live a healthy lifestyle and encourage you to do the same.
- Friends who call you on your manure.
- Friends who have similar values and beliefs.
- Friends who don't judge you when you make mistakes.

- Friends who see your superpowers, especially when you can't see them yourself.
- Friends who have an incorruptible integrity.
- Friends who tell you what you need to hear and not just what you want to hear.
- Friends who exert positive energy and have a positive attitude.
- Friends who are adaptable to change.
- Friends who say what they mean, mean what they say, and do what they say.
- Friends who will tell you if you're headed in a good direction, or be straight up with you when you're not.
- Friends who help you to get back on the horse when you fall off.
- Friends who have faith and encourage you to keep yours.

- Friends who are funny—both ha-ha and quirky.
- Friends who are there for you through thick and thin.
- Friends who motivate and inspire you to serve others.

Of course, I have fond memories of people I grew up and went to school with. Many of us still keep in touch. And if I sense someone is coming over to the dark side, I get neighborly and check on them.

However, as I've grown older, I'm more selective when choosing my herd. Nowadays, I'm perfectly okay not being connected to anybody and everybody. My faith and confidence are much higher—my friendships are now about quality over quantity.

As you grow older, you will find out who your friends are, too.

♫ You want somebody who's gonna drop everything, who will run out and crank up their car. Someone who never stops to think, "What's in it for me?" or "It's way too far." They just show up with their big old heart. ♫

Great, now I've got that Tracy Lawrence song stuck in my head.

My friend Dardi is good stock. She has highly sought-after qualities like good health, nurturing abilities, adaptability, temperament, and behavior. In fact, she's so good, she'll tell you, "You can call me anytime, day or night. You just can't call me from jail."

Use the qualities listed above as your benchmark and refuse to settle for less. This works both ways, however—refuse to be less than these qualities yourself, as well. Learn to possess those highly desirable characteristics so set an example and pass them down to others. If you sense that someone is in trouble, be neighborly and reach out to them. Always be a good steward to others.

Your herd should boost your faith, happiness, and living conditions. Don't be afraid to cull those who don't. When you build a great herd, you build immunity for the dark side of college and life.

And if I can find my herd, so can you.

Real friends

~Friends comfort you with comforting words. *Real* friends comfort you with comforting words too, but with sarcasm.

~Only your *real* friends tell you when your butt is too big.

~Friends give you a shoulder to cry on. *Real* friends are ready with a shovel to hurt the person who made you cry.

~You can always tell a *real* friend: when you've made a fool of yourself, they don't feel you've done a permanent job.

~A *real* friend laughs at your jokes, even though she's already herd them all.

PART 3

HARVEST YOUR SUCCESS

Plant your seeds, Grow cash crops

Believe To Receive

"If you think you can do a thing or think you can't do a thing, you're right."

Henry Ford

I was on a coaching call with a client who hadn't had much luck finding a new job. So, I wanted to get an idea about what he'd been doing search-wise and what his thought process was.

I asked, "So far, what has been your strategy?"

"Prepare for the worst and hope for the best," he replied.

"That's not a strategy; that's the kiss of death!"

Our minds have the power to make or break us. If you believe you'll get a great roommate, teacher, or class, then that's what you'll get. The same is true for the opposite. If you dread particular classes or, in this case, a job search, it will be dreadful.

When I was getting my master's, I started out in a degree plan that turned out to be a poor fit for me. However, the more I moaned and groaned about it, the more I felt stuck. The more stuck I felt, the crabbier I got. The crabbier I got, the more my creativity dried up. And just like the old shampoo commercial says: and so on, and so on, and so on (if you don't get it, look it up).

Luckily, you have the power to shift gears and change your thinking. You can turn your situation around. It all starts with creating a vision.

By now, you know your superpowers, beliefs, and values. With this information, you're light-years ahead of others. So, now it's time to come up with a plan. After all, you wouldn't build a house without a plan; you can't build a thriving life without one, either.

Creating your vision

When you understand yourself, it's a whole lot easier to write a vision. Below are the steps that will help you create your vision.

1. Figure out your superpowers, beliefs, and values. If you need a reminder of what these are, revisit chapter 2.
2. Decide and list what you want, no matter how far-fetched.
3. Get it down on paper. Something magical happens when your thoughts come out of your head, down through your hand, and out on the paper.

4. When you're writing it out, sometimes it helps if you start with your fears or limiting beliefs.
5. Then flip the script!
6. Use Afformations (different from affirmations) to state what you want in a question so your brain tries to solve the problem. (this is a good 'un, look it up).
7. Recite it three times a day, so the good stuff simmers in your brain.
8. Allow the universe to work its magic, and extraordinary things will start to happen.

Let's sprout some more ideas to help you "flip the script":

Instead of thinking and saying: This school stuff is too tough for me.

You could say: Why do I find a degree plan or skills I enjoy and is a good fit for me?

Instead of thinking and saying: I'll never make friends as I did in high school.

You could say: Why do I find new real friends who are supportive, faith-focused, and like-minded for my new herd?

Instead of thinking and saying: I'll never get through my statistics class.

You could say: Why do I have a great teacher who explains it well, and I pass with flying colors?

Instead of thinking and saying: I'll never figure out a good direction for me.

You could say: Why do I work with an awesome coach who helps me discover everything good about me and what I find important, so I can be successful?

You may not be in the market for a career just yet, but this will still help you create a **career vision** now.

Use the acronym CAREER to write the vision. Here are some ideas on what to include.

- **C**ulture: *describe the culture you want to be in.* For instance, ethical, with a work-life balance, humorous, people treated as assets, socially responsible, and having aligned values.
- **A**uthentic Leadership: *describe the type of leaders you want to work under.* For instance, fair, hands-off, laid-back, trustworthy, compassionate, and a leader who's my advocate.

- **R**elationships: *describe relationships with coworkers, colleagues, and clients.* For example, respectful, honest, fun, and collaborative.

- **E**ngagement: *describe what it takes to make you excited about going to work.* For example, using strengths, skills, and passions and duties that fuel my exhilaration.

- **E**ducation: *describe what you want to learn to make you more valuable in your career.* For example, MS Suites, G-Suites, and coding. And also, if continued education is encouraged and provided at the employer's expense.

- **R**ewards: *describe how you want to be rewarded both inside and out.* For instance, serving others for a higher purpose, connection, and exhilaration for the inside. And for outside, your pocketbook being $85,000 per year, insurance, holidays, vacation, and sick leave (not combined), education reimbursement, paid time off (PTO), stock options, and 401K.

Again, use afformations (different from affirmations) to state what you want in a question so your brain tries to solve the problem (friendly reminder to look it up).

Instead of thinking and saying: I hate my job!

You could say: Why does this job teach me new skills, help me build relationships I can use in the future, and prepare me for a better-fitting job?

Instead of thinking and saying: I'll never find a good place to work.

You could say: Why do I end up in a culture where we laugh all day, the company trusts the employees, and they treat us like their greatest asset?

Instead of thinking and saying: my boss is going to micro-manage me to death.

You could say: Why do I end up with a boss who's hands-off, laid-back, and supports my professional growth?

Instead of thinking and saying: I won't find a role that makes me happy or pays me well.

You could say: Why do I get to use my superpowers as much as humanly possible, I so happy that I can't wait to go to work every day, and my company sees my value and pays me my worth?

Remember you'll want to keep this book handy later so you can later reread how to **create a career vision** *Anyhoo*, I hope you get the picture.

Oh, by the way, if you're gonna create a vision, you might as well make it the ideal situation. This can be scary. I mean, heaven forbid we get the kind of life we really want!

Once you get your ideas and dreams outta your head and onto paper, something magical happens. Your superpowers start working on the plan, turning your vision into reality.

As Paul Harvey would say: "and now...for the *rest* of the (college) story..."

After I stopped wallowing in my misery and flipped the script and, I began researching different leadership programs in Texas. I looked at many schools I hoped were a good culture fit for me. I knew in my gut that leadership was too generic of a degree. Then all of a sudden, I began to run across agricultural leadership programs. That direction felt right, so I went with it!

Unfortunately, I wasn't finding any online programs to my liking. My gut kept telling me to contact Tarleton State University, but of course, I blew it off for a while. I came up with all sorts of excuses: they won't have anything online, they aren't gonna have what I'm looking for, this vision stuff *never, ever* works, etc. Finally, that gut feeling became so strong I couldn't resist it no mo'. It felt like somebody was smacking me upside the head with a 2 x 4.

And so, I went to their website. Low and behold, just like I had written in my vision, they offered an online master's program in agricultural leadership that didn't require the GRE. I felt like I'd hit the jackpot!

On top of that, I had great teachers. One of them even got me through stats with flying colors! I graduated with my master's from a Texas school with a degree that felt right for me, that I'm proud to be a part of. It was aligned with my vision. I'm now doing what sets my heart on fire in an industry I love and wholeheartedly respect.

And now that you know the rest of the story: hope for the best and prepare for the best! Create your ideal vision for college and your career, then believe to receive. Watch those superpowers kick in and allow something extortionary to happen. Do this for yourself and for the good of others.

And if I can create, believe and then receive, so can you.

When I was in school, I thought I had special powers.

All my teachers called it something like "constant super-vision".

Work Your Faith Muscle

I grew up in the Methodist church. Reverend Zirkle told us happy stories about Jesus and the plan He has for us. He told us to share God's love, prosper, have hope, and have a future. We sang some all-time favorite hits like *How Great Thou Art*, *I Love to Tell the Story*, and *Jesus Loves the Little Children*; I sang at the top of my lungs!

We also had pins for Sunday school. For every year of perfect attendance, you'd get a new bar to add to the pin. Since I got perfect attendance eight years straight, my pin made me look like an Army General. Moreover, I loved going to church camp, and Youth Fellowship every Sunday evening. Heck, I even liked the confirmation classes!

As you can see, my church-going memories were happy ones. For my husband, however, this wasn't the case. He was raised Pentecostal. When he started rattling off Bible stories involving beheadings, plagues, and Armageddon, I looked at him confused and said, "I thought Armageddon was just a Def Leppard song." (If you don't get it, look it up)

Despite my happy memories, I still lost my faith and my way. My life was going quite poorly. When it came to friends and men, I knew how to pick 'em. I hung out with and dated some colorful people. But that's not all. One time, I ended up with a broken arm. I started thinking everything I learned in Sunday school was a bunch of horse hockey. I blamed God. How on earth does He let bad things happen to good people? I'm good people! And if He is so freakin' powerful, why doesn't He fix my life already?

Of course, it's much easier to blame God than take personal responsibility, so I got stuck in a vicious cycle. The less faith I had, the more my life didn't work. The more my life didn't work, the less faith I had. And so on, and so on, and so on.

At one point, I hit rock bottom (turns out, one of many). I had no plan, little hope for my future, and continued making lousy choices—drinkin', eating, and spendin'. My depression and anxiety were high and my confidence and happiness were low. I kept going down deep, dark, insane rabbit holes, and I couldn't pull myself out.

I wanted to know why those bad things kept happening—why those happy Bible promises weren't coming true. So, I decided to pay Reverend Zirkle a little visit. I don't remember the entire conversation, but he reinforced His plan, encouraged me to believe in somethin' more, and assured me that

♫ I gotta have faith—ooh I gotta have faith, faith, faith. ♫

Great, now I've got that George Michael's song stuck in my head (look it up).

I felt better because my faith and sanity were restored. For a little while, anyway.

Unfortunately, for many years, I struggled to believe in somethin' more. I couldn't see that Something or Someone really wanted great things for me. I especially didn't feel like I was being cared for along the way.

And during this struggle, my life continued to go poorly. I made more lousy choices in both friends and men. On top of it, I was always broke. I chose jobs that were not a good fit, I indulged in many unhealthy hobbies—spendin', drinkin', and eatin'. While some days I treated my body like a temple, most days it was in ruins.

Eventually, I stumbled into a 12-step program (literally). Although *my* life wasn't working, I told myself this was because of everybody else. So, I came to fix *them*. In my first meeting, I swore it was a cult. But I stayed "just to see" if I could get *them* to behave, so my life might get better.

While I came in convinced that my external circumstances were the problem, the fellowship suggested that my issues were internal. And to recover from my crappy life and be restored to a happy, saner life, my faith (or lack thereof) needed to be addressed.

However, many members, like myself, were fed up with all this talk about faith and religion. We didn't need no stinkin' God or religion! Look where it had gotten us so far! So, even though the fellowship was established by religious founders based on religious beliefs, the founders decided to tread lightly. As such, they referred to the fellowship as a spiritual program to avoid skeering anybody off.

Some people get downright offended when you don't use the word God while others get offended when you do. Can't win for losing. These days, I flip-flop referring to Them as God or the Universe. I don't believe the name matters. What does matter is that you create a positive, trusting relationship with a power greater than yourself. What matters is that you're willing to turn over your life, struggles, and outcomes to Them. And that you practice, practice, practice building that faith muscle. What matters is that you understand that when you do this, your life will have fewer struggles and better outcomes than when you try to take the reins all by your lonesome.

On my spiritual journey, there were a few tips I learned that helped restore my faith and my sanity:

TIP

No matter the addict's drug of choice, there are two common denominators: a core belief you ain't worth a hill of beans and a divine disconnection.

Most who struggle with the God concept carry those beliefs. However, if you don't address them, nothing changes.

If this is you, chances are, there was someone in your life who might have been critical, punishing, or acted holier-than-thou? This caused you to associate that person with "the real McCoy," and those characteristics became the God of *your* understanding. And at some point, you lost your way and your faith because those characteristics made it difficult to trust them, yourself, or anybody else.

In this case, it might be time to find a new God of your understanding. Give Him, Her, or whatever you want to call Them some positive qualities—maybe use that list you made of your favorite person! Describe Them the way you would like to see Them. Then, describe the relationship you would like to have with Them and build it. With this, you can begin to trust Them, yourself, and others who have your best interest at heart. By growing the relationship, you can rely on Something or Someone greater than yourself. Just remember to make Them a good 'un because Them is *always* good, all the time.

On another side note, if you had a person in your life with those *other* not-so-great characteristics, I can assure you they, too, carried those "common denominators" throughout most of their life. Decide to forgive them. Focus on your own life, so you have peace of mind and peace of heart. Allow the good to come flowing through.

Faith is a state of mind that gives you the ability to believe in Something or Someone more without being able to see it.

The struggles we have are usually of our own making.

However, God, Source, Higher Power, the Universe, Infinite Intelligence—whatever you want to call Them—is *always* good, all the time. If you hold still long enough, your struggles will lessen, and you will feel peace of mind and peace of heart. This allows the good to come flowing through you.

Faith keeps you aware of your surroundings while helping you hold an abundant mindset.

Despite what others might say about the ongoing in our world, faith tells us we are surrounded by good, all the time—good stewards, generosity, fellowship, food, nature, and prosperity. When you believe more of the good stuff is headed your way, it will. And you'll be cared for along the way.

Faith gives you the confidence and ability to defer instant gratification.

If you can wait a little while longer, you may end up with something bigger and better than you could have ever imagined.

Faith helps you practice taking personal responsibility.

The more you create that positive, trusting relationship with Them, the more willing you become to see the part you played in your circumstances. Believe you me, everyone can create their own happiness or crappiness.

Faith gives you confidence and maturity.

A preacher once said, spiritual maturity is the adamant refusal to draw attention to yourself. I'm not gonna sugarcoat this: we live in self-absorbed, selfie-posting world, and it drives me bat crap crazy. Recently, I took my first selfie in five years. The only reason I took it was for a contest. I had a deadline, my husband, Maverick, was out running around on the property, and the prize was pretty *saweet*.

I felt so uncomfortable taking the selfie that I added my *way* cuter-than-me dog, Duke, in the picture. It took me a while to take the silly thing. By the time I figured it out, Duke had said *screw it*, laid down, and took a catnap.

Having faith gives you confidence in yourself. It helps you act maturely. Don't do anything for recognition, like posting selfies to fish for compliments. Instead, build your faith muscle and spiritual maturity. Be as adamant about doing this as you would be about collecting likes, comments, and shares.

TIP

Faith teaches you to appreciate what you have.

Appreciation means you have the ability to recognize, take pleasure in, and enjoy the abundance in your life. Again, when you focus on the good stuff, you get more of it. So, pay attention because we are surrounded by lots of goodness.

TIP

Faith guides your core set of beliefs, principles, and values, which helps you fully serve others.

It will help you understand and align with what matters to you. Faith gives you the courage and confidence to make good choices that are a good fit for you. And it helps you stick to your guns in the process. When you take what you understand about yourself and use your powers for good, there's no stopping you.

Dog is God spelled backwards.

Somehow, I ended up an avid collector of the creatures. I provided them with a good home and saved a few lives. In turn, they saved my sanity. *Coinkidink?* I think not!

The more you exercise your faith muscle, the stronger it becomes.

The stronger it becomes, the more perseverance you have. The more perseverance you have, the more you get stuff done. The more you get stuff done, the quicker miracles happen. The quicker miracles happen, the greater your faith becomes. This is a very virtuous cycle.

Of course, there will be days when your faith is tested and make you scream at the sky. For me, the hardest thing to do on those days is to muster up enough courage to trust the timing of the process. To be honest, when things are challenging, I can still have my pity party and struggle to turn over my life and the outcomes-I turn it over, then take it back, turn it over, then take it back... This is human nature, however, so I try not to clobber myself in the process.

What matters most is that we don't lose faith. Appreciate what you have. Create a positive, trusting relationship with Something or Someone greater than yourself. Know you're being cared for. Trust the timing because Them is *always* good, all the time.

Work your faith muscle but let Someone else do the heavy-lifting. Build your spiritual maturity. Serve others. Do good for no reason other than for the good of others. There's a plan for you. You will prosper. You will have hope. You will have a future.

♫ You gotta have faith. Ooh, you gotta have faith, faith, faith. ♫

And if I can continue to work my faith muscle, so can you.

Faith **is like Wi-Fi**

It's invisible but it has the power to connect you to what you need.

Ka-Ching, Ka-Ching, Ka-Ching!

In case you didn't already know, the term cash cow is a metaphor for a dairy cow that produces milk over the course of its life, again, and again. While the term is commonly used to describe profitable business ventures, it can also be applied to your career. I'm assuming after you finish school (no matter what road you've taken), your next step is to find a career that makes you happy and produces these cash cow returns:

- A strong, steady, reliable cash flow.

- The ability to reinvest in other areas due to the abundant income.
- High rewards.
- Funds for your growth.
- Leverage in the pursuit of other high-return opportunities.
- Positive effects to spill over to other areas in your life.

I can assure you that for your career, you'll compete for roles, bonuses, customers, awards, and promotions. To yield cash cow results, you'll want to take the game-changing qualities and practices you cultivated as a student and implement them into a professional brand (remember chapter 3?). Your job is to create a brand that makes you stand out from all other candidates. Your brand should make you distinct and recognizable to leaders, coworkers, colleagues, and customers. Most importantly, your brand should include being of service to others. When you set yourself up to be a successful professional, you will more than likely end up with a cash cow career! This is what I did.

As I got older, I discovered my strengths, skills, and passions and recognized I wanted to utilize my superpowers as much as humanly possible in my career. I wanted to promote causes I believed in and was passionate about. This meant working for companies with similar values. To accomplish this, I started to stand out in my field and, as a result, was recognized by leaders, coworkers, colleagues, and customers as distinct and recognizable. In my newfound awareness, my life began to change—I began to have cash cow careers.

Doggy drug dealer

I was so excited when I figured out a role as a veterinary sales rep was a great fit for me! I had the strengths, skills, and passion required, and it felt like the perfect fit.

Then, the naysayers came out of nowhere like baseball players coming out of a cornfield. Some of them actually said to me:

"Do you know whom you have to know to get a job like that?"

"Why would you even want a job like that?"

"Do you know how long it takes to get a job like that?"

Even though the direction felt right, I got sucked into their fears. I second-guessed myself. This caused my application process, myself, and my results to be scattered.

When I saw what this trickle effect was doing to me, I stopped myself and regrouped.

I decided I had enough naysayers going on in my own head—listening to them wasn't doing me any good, and I was burning daylight! I decided it was in my best interest to steer clear of that herd.

So, instead of wasting time and energy focusing on what they feared, I created a cash cow career vision. I listed everything that mattered to me, no matter how far-fetched, and I recited it three times a day.

When I did this, magical things began to appear out of thin air. You should have seen the synchronicity! The right people showed up when I least expected them. A new connection in the agricultural industry suggested I attend a veterinary conference to get in front of hiring managers. I found the right conference, which took place that same week.

I truly wanted this role, so I came prepared. I studied the list of exhibitors, researched the companies to make sure they were a good fit, and found the names of their hiring sales managers. To avoid wasting time, I mapped out their exhibitor's booths, so I knew where to go when I entered the hall.

I had all the right conversations. On the two-hour drive to the conference, I'd practiced my elevator pitch. At the conference itself, I initiated conversations with hiring managers. Not only that, but I handed out fifteen resumes. As a result, I had four sit-down interviews on the spot.

The right place at the right time occurred. The following week, I had my ideal job at a Doggy Drug Rep, with an offer that included 98% of my vision. *Woo doggy!* Best part, I got to love on all kinds of critters and serve others.

And I stood out from the herd: I showed up when I was scheduled, didn't call in to bail, and kept my attitude and energy positive. I took ownership in their needs and bent over backwards to solve their problems. I didn't fear the competition, I made friends with them so my customers got the products they needed, when they needed them. As a result, veterinarians and practice managers, even my colleagues out in the field, felt my passion, genuine interest, and trusted me. I turned my passions and partnerships into profits and this gig turned into a cash cow career.

Lesson learned: steer clear of the naysayers, especially those in your own head. Write a vision and commit to it. Create positive associations with others, increase desired behavior with positive reinforcement, and stay consistent.

And if anybody asks if you know how long it can take to get a role like that, tell them seven days.

Career readiness instructor

At this point in my life, I created a vision for having a career services role within a trade school in a specific location. As before, I recited it three times a day, every day. Within a couple of months, I got hired by the right trade school, in the right location, and at just the right time. What wasn't right, however, was the role I took. Still, I figured three out of four ain't bad. *Boy*, was I wrong.

Instead of sticking to my career services vision, I took a high-pressure sales job as an admissions rep out of fear and desperation. First lesson learned here: nothing good comes from making a decision from a place of fear and desperation.

Well, almost nothing.

The part I loved about this job was helping potential students find their way. However, I loathed the pressure to be on the phone every hour of the workday while my boss sat three feet away listening in on my calls. I was even too scared to take bathroom breaks. The role was so wrong that I started to get sick from kidney infections. In addition, my depression and dread became so high that I could hardly pull myself out of bed each day. And at night, I cried myself to sleep.

However, since I've got thick skin and a workhorse's work ethic, I made a commitment to myself to tough it out in admissions for my 90-day probation. During that time, I took the opportunity to plant the seed of interest with the career services director for a career services role. Unfortunately, he had just hired a black, veteran woman. Even as a woman myself, that's a lot of diversity to be up against. Still, he loved my passion and energy and knew I had built trusting relationships with my leaders, coworkers, and students. Nevertheless, he also assured me that he saw nothing on the horizon in his department.

And so, I also planted the seed of interest with the head of faculty. Unfortunately, we were in the process of moving into a new million-dollar building. Although she, too, loved my passion, energy, and the trust I'd built, she assured me the number of students wasn't high enough yet to support a teaching role for me.

It felt like nothing was working out, and my faith was fading fast. However, even though my current role was not a good fit, I still believed that my vision was what the universe wanted for me. And so, through tears of misery, I stayed committed to my vision and continued to recite it three times a day.

As I came to the end of my 90-day probation period, I took a trip to the beach to regain some sanity. Sobbing uncontrollably at my happy place made it clear to me I couldn't stand being in my current role any longer. I remember looking up at the sky and screaming, "What the heck are you doing to me?"

I knew it was time to strap on my boots and do what was right for me. I decided to go in on Monday and give my two-weeks' notice. Even though I had nothing else on the horizon, I believed in my vision so strongly that I knew it would come, somehow, someway. After all, I was living proof of how well things had worked out at other times.

So, on Monday morning, my resignation letter was printed and ready to be handed in. I went to the copier to make a copy for my boss. And just as the copy fell into the tray, the director of career services rushed in to tell me that the veteran woman he hired was leaving to work with other veterans. He offered me the career services role on the spot. Talk about the right timing!

That opportunity turned into a teaching role, where I used all my superpowers and taught students to discover theirs. We covered everything from cover letter to offer letter—resume writing and tailoring, interview skills, LinkedIn, salary negotiations, job searches, customer service, and professionalism. I had found the right role for me.

Best part: the students were motivated as I taught students how to find the right role for them. Most were very successful. The faculty, admission reps, and students associated me with trust and positive energy. I was able to help everyone serve others.

Lessons learned: commit to your vision. Don't settle for less, at least not for long. Trust the timing. Have faith, the universe has great things in store for you.

Career coaching

Eventually, management changed at the trade school. I know myself well enough to know micro-managing is not something I respond well to. This meant the role was sadly no longer a good fit for me.

I left to work for our local Workforce Solutions, which is the other side of unemployment. I loved helping job seekers with their resumes, interviewing skills, and job hunting. So, this job was a great fit.

Unfortunately, I quickly learned I was trying to help people find work who didn't really want to work. Since I have a workhorse's work ethic, I realized neither the customers nor the company I worked for shared my values.

However, just because the job lost its fit doesn't mean all was lost. I took advantage of my time there. I strengthened my coaching, resume writing and interviewing skills, and job-search knowledge. It funded my growth. In addition, I got certified as a career coach and interviewing specialist. Best part, I got to serve others.

Again, I wrote a vision and recited it three times a day. A few months prior to leaving, I had an interview with a career transition company, but they assured me there were no roles available. When I finally lost all desire to work at Workforce, I decided to leave to walk dogs instead. With my background, I knew customers would trust me with their babies. In addition, I would make more money, and nobody would scream at me.

With that plan in mind, I gave my two weeks' notice. TWO HOURS LATER, the career transition company called to offer me a career coaching role. That's got to be warp speed! When they sent me my offer, I compared it to my vision. The wording was so close; it was as if they had plagiarized it.

For more than three years, job seekers and I spoke one-on-one about everything job-related. We also talked about family, the importance of health and wellness, stress management, career vision, and fears. Some days, part of serving others meant talking them off the ledge.

Each day, my exhilaration skyrocketed, and I couldn't wait to jump out of bed to get to work. Oh, by the way, in the vision I wrote, I work as close to home as possible. This role allowed me to work *from* home. It doesn't get much closer than that.

Clients associated me with being prepared, trusting, a good listener, motivational, and empathetic. They related to my colorful career stories and lessons. We connected easily because I had been one of them.

This role produced many cash cow career benefits:

- I strengthened my coaching skills and helped my clients become game changers in their job search and professions.

- It brought high rewards—both inside my heart and outside in my pocketbook.
- It allowed me to fund my growth. I became a certified executive coach and finished my master's degree in agricultural leadership.

I felt *absolute* career happiness and often referred to this role as my dream job. Those positive effects spilled over to other areas of my life—my creativity, health and wellness, and relationships with others and myself. When clients implemented their new beliefs and practices and saw a good return on their time investment, they trusted me enough to help out their friends. Best part: over 600 of clients found work that made them happy and ended up with cash cow careers. Their success was always my happiness. I got to serve others.

Lesson learned: write out your vision and commit to it. Not only do what makes you happy, do what makes you feel exhilarated. Serve others so they can be successful.

Academic advisor

My career coaching role gave me another cash cow benefit—with my experience, I now had leverage to pursue other high-return opportunities. While I was coaching, I decided I wanted an even stronger cash flow. A role as an academic advisor felt like a great complement to what I was doing. As before, I wrote a vision and recited it three times a day.

On a Monday, I applied for the only advising job at the only school I had in my vision. Two days later, I got a call from that school to schedule an interview for the following Monday. I interviewed for 30 minutes, left, and within an hour, I was offered the advising role.

In this role, the students, their parents, and other advisors associated me with trust, clarity, and a *You Got This!* attitude. They referred me to their friends, and students kept coming back to me each semester. That cash cow career strengthened my advising and coaching skills and provided a steady, reliable, abundant cash flow. I received high internal rewards because I was able to be a good steward to students, and their success was always my happiness. I was able to fund my growth, and because of that, I now have more leverage to pursue other higher-return and *higher-purpose* opportunities.

The positive effects spilled over to other areas of my life—my passion for helping students to be successful exploded, my creativity kicked into high gear, and my college knowledge poured out of me. That's how I wrote this book specifically for you. Best part, I got to serve others.

Lesson learned: when we do things that make us come alive within our careers, we thrive. In that state, we tend to treat ourselves with more appreciation, acceptance, compassion, and kindness. This has a trickle effect on the way we treat and serve others. Oh, and write out your vision and commit to it. End of story.

How to create a cash cow career

Know Yourself.

When you're willing to understand yourself, you'll discover your strengths, skills, passions, and values. Use this information to your advantage when you're searching for a career that makes you happy and produces cash cow returns. You'll begin to recognize roles that utilize your superpowers. In addition, you'll begin to spot companies that share your values. Best part: you'll know yourself well enough so that you won't want to settle for less—at least not for long.

During interviews for roles and companies that are a good fit, you have to tell them about yourself. If you know yourself, you can do this with confidence. You'll focus on your strengths, skills, passions, and successes. Best part: your answer is focused on how you'll serve them so they can be successful. People hire those they like, and when you make it all about them, they tend to like you *a lot*. Remember: it's all about them.

Understanding yourself and creating a vision will give you the courage to go for those cash cow careers. Make decisions based on your superpowers and are aligned with your vision. When you do, you'll explode with positive energy.

I've had a lot of cash cow careers over the years. As a result, my resume is a colorful combination of stories and lessons. It has been a process of elimination, and some days termination. Regardless, each job strengthened my natural abilities and instincts. Each job helped me strengthen my professional brand. Each job prepared me for the next one.

Cultivating a professional brand is about taking ownership of yourself, so you can take ownership of the needs of others and solve their problems. Know your strengths, skills, and passions. Know your values and be willing to stick to your guns. Implement some game-changing practices and serve others to help them be successful.

Carry the awesomeness you practiced as a student into your profession. Remember, people hire those they like, so have your manure together and build their trust. Take initiative. Be prepared. Take ownership.

Have the courage and confidence to do what sets your heart on fire and turn your passions into profits, but make it your sole purpose is to serve them. When you create this combo, you'll be rewarded inside and out. Your heart and your pocketbook will both be singing *ka-ching, ka-ching, ka-ching!*

And if I can create a cash cow career-or 22 of them-so can you.

Ag-Related Job Titles

If you wonder what kind of agriculture career you can have when you graduate, here are some titles to noodle around. Begin NOW to research the titles that interest you. Know what someone does in that role and how much it pays. Then, reach out to people in those roles to pick their brain about what to expect. Ask to do a ride along to find out what a day in the office *really* looks like. Do not be afraid to test the waters!

Make sure that line of work will be a good fit for you.

All of these titles can be used in the agriculture industry

Agricultural and natural resource production specialists

- Arboriculturist
- Aquaculturist
- Farmer
- Feedlot manager
- Fruit and vegetable grower
- Greenhouse manager
- Farm manager
- Landscape designer
- Landscape manager
- Nursery products grower
- Pest manager
- Ranch Management
- Turf manager

- Turf producer
- Viticulturist
- Wildlife manager

Social services professionals

- Career counselor in a college agriculture department
- Community development specialist
- Conservation officer
- Consumer counselor
- Environmental lawyer
- Extension Agent
- Food inspector
- Labor relations specialist
- Natural resource policy analyst
- Naturalist
- Park manager
- Peace Corps representative

- Politician
- Regional planner
- Regulatory agent
- Youth program director

Technical service representatives

- Communication and education specialists
- College instructor or professor
- College student services administrator
- Conference manager
- Consumer information manager
- Conservationist
- Cooperative extension agent
- Editor
- Educational specialist
- Government lobbyist
- High school teacher

- Illustrator
- Information specialist
- Information systems analyst
- Journalist
- Natural resource interpreter
- Outdoor recreation specialist
- Personnel development specialist
- Public relations representative
- Training manager
- Writer

Financial and marketing specialists

- Accountant
- Account executive
- Advertising manager
- Appraiser
- Auditor

- Banker
- Business manager or owner
- Commodity broker
- Consultant
- Contract manager
- Credit analyst
- Customer service manager
- Economist
- Export sales manager
- Financial analyst
- Food broker
- Government program manager
- Grain merchandiser
- Grants manager
- Human resource development manager
- Insurance agency manager

- Insurance agent
- Insurance risk manager
- Landscape contractor
- Market analyst
- Marketing manager
- Purchasing manager
- Real estate broker
- Research and development manager
- Retail or wholesale manager
- Sales representative

Scientists

- Animal scientist
- Bioresource engineer
- Cell biologist
- Dentist
- Ecologist

- Environmental scientist
- Food scientist
- Fire scientist
- Fisheries biologist
- Geneticist
- Marine biologist
- Medical doctor
- Microbiologist
- Molecular biologist
- Natural resources scientist
- Pathologist
- Physiologist
- Plant scientist
- Quality assurance specialist
- Rangeland scientist
- Research technician

- Resource economist
- Soil scientist
- Toxicologist
- Veterinarian
- Waste management specialist
- Water quality specialist
- Wildlife biologist

Credit: Oregon State University

The Cowgirl Rides Away

As I told you before, I love working with ag kiddos, especially because they work hard and are tough. Ag people are my people.

I wrote *Don't Farm Naked* to give you tips, stories, and lessons to help you harvest success. No matter how humiliating an experience I cultivated for myself. Before this cowgirl rides away—probably off to watch Yellowstone--I want to leave you with some takeaways from the book:

Take the right trail. Have a plan so you feel great about your future.

Know yourself. Go with your natural grain, not against it. Use your superpowers for good—both your own and that of others.

Build your brand. Be intentional and strategic. Be a game-changer in college and in your career. Stand out from the herd.

Learn to find information yourself. It will empower you.

Find the right educational route for you. Set yourself up to be future-proof.

Match your personality with your classes. Also, find out if they transfer by speaking with an academic advisor.

Nothing in life is actually free. Be smart about taking on debt. Focus on finding school money you don't have to repay because paying it back isn't "optional."

Know that sometimes life can get lonely, so surround yourself with a herd of *real* friends. Find friends with highly desirable characteristics, and don't be afraid to cull those who don't. You want friends who see greatness within you, even when you don't see it yourself. And remember: be the same kind of friend to them.

Write down your ideal vision, no matter how far-fetched. Believe in your vision. Magical things WILL happen.

Work your faith muscle. Build spiritual maturity. Do something for no other reason than because it benefits others. Trust that you will be cared for along the way. Know there is a plan for you. You will prosper, have hope, and have an *Ah-Mazing* future.

Have the courage to go for careers that make you happy. Pursue higher-return and *higher-purpose* opportunities. Turn your passions into profits. Make it your sole ambition to help others become successful.

Don't farm naked.

As country star Reba McIntyre said:

To thrive in life, you need
three kinds of bones:
A wishbone; a backbone; and a funny
bone—my favorite part of the bird.

If I wasn't willing to poke fun at myself, I'd lose my ever-loving mind. As you've read, I've done it all wrong...and then right. I've shared my colorful college and career experiences so you'll know how to do things right from the start.

I hope this book gave you insight on what to expect, motivation to get back on the horse when you fall off, and a grin from ear to ear. I want you to harvest TONS of success. And not just that: I want you to do things that make you come alive and to get to the place you were meant to be: living a thriving life. Your success is my happiness.

And if I don't farm naked, neither should you.

About The Author

Donna Fuller uses her superpowers for good as a funny, motivational speaker, academic advisor, and career coach.

Although her college and career experiences are quite *colorful*, her passion for helping students and professionals is unmatched. With her good, clean humor, she educates and entertains others so they yield a variety of highly successful careers.

As an academic advisor and executive career coach, Donna has helped HUNDREDS of students and professionals get on the right trail, create a plan that puts them in the right direction, and get them in a cash cow career! After working with Donna, her people feel clear, hopeful, and empowered.

She has won several humor awards for her speeches and stand-up comedy routines. Her first book, *Happy Work. Happy Life. An Inside Job.* got 5-star reviews, and not just from her mom.

Donna and her husband, Maverick, live in a barn-do-minium on 45 acres in Texas, with all their critters.

Contact Donna

Donna coaches and speaks humorously on having college success and career happiness. She is available for keynote, half-day, or breakout sessions covering the content of this book. Donna will educate, entertain, and empower your audience. If interested in finding out more, please contact her at:

www.FullerAgCoach.com

email: dfuller@fulleragcoach.com

About Zane-The Model

Zane Cowan graduated from Canyon High School in New Braunfels, Texas. His entire life, he immersed himself in agriculture--everything from FFA to working summer jobs at various ranches to being a lifetime member of the San Antonio Stock Show and Rodeo. To top things off, he is also an Eagle Scout.

Currently, Zane is a student at Tarleton State University majoring in Animal Science with a concentration in Range and Ranch Management.

And now that I have disclosed Zane's whereabouts, I'm pretty sure Tarleton's admissions rate will skyrocket.

Acknowledgments

Thanks to Jim Bob Solsbery for being my ag voice of reason, especially when I needed talking off the ledge.

Thanks to Dr. Frances McArthur for being my teacher voice of reason, especially when I had doubts about this project panning out.

Thanks to Iris Marsh and Suzanna Brown for your editing and proofreading expertise so this book flowed properly.

Thanks to Zane Cowan for being my cover model. You are such a trooper!

Thanks to my family and friends who supported me while I cranked this out.

Thanks to every student and parent who reads this book. I hope it gets you to the place you were meant to be.

Thank you for letting me serve you.